Really!
No-Body Knows This!
Are These The Actual Answers To Global Warming?

From a 9 Year Olds mind of Scientific Intuition comes:

A Small But Powerful Compendium of:

- o Self - Discoveries With Incredible Possibilities
- o Who, Where, What, Why and the Big YES!! Solutions into the future.

You Can Just Call Me: A.J. The Global Warming Herald

Protected with a Registered Copyright through the U.S Copyright Office

Page Intentionally Left Blank

Notes:

_____

_____

_____

_____

_____

_____

_____

_____

Table of Contents:

Contact Me Here for Anything

TheGlobalWarmingHerald@Gmail.com

Hey it's me, A.J. My Introduction:

"The first draft of anything is always shit" Ernest

Hemingway

One night in a dream I had a conglomeration of visions that lead me into the start of this book. They say that Thomas Edison discovered many of his inventions while in a dream state of mind. There may be something real to that, hmmm, what if I used this to help the world - Just like Thomas did.

There are many approaches I could have used to tell the world about my dreams and visions. I started with an emotional approach, making people mad or worried, but then I said to myself - *"That is what the media is for, isn't it?"* Then there was the Social Media method where you create a viral set of content all geared to emotional attention. *"Nahh, that's not going to be good A.J" I said to myself.* Like Ernest said the first draft of anything is always shit. What if by this, he didn't mean grammar or literature structure. What if Mr. Hemingway meant the first draft is a superficial perception, and as you write more and more drafts, you peel away until you get to the REAL CORE of what you are saying. What if each draft is just the removal of a layer of skin belonging to your EGO. "Yeah, that's it. That is what you meant Ernie". I said to myself. These are the thoughts that lead me to this books indifferent approach to the norm.

I would like to invite you all into my minds pleather of perceptions. Ohh, my English teacher would love the way I said that. "Big words A.J., Big words he would say" Enter as I take you through an incredible amount of new thoughts on where science might be missing the target, overlooking real causes not yet spoken about (At least to my knowledge) but also offering up a complete digestion of ideas, knowledge, hope, perceptions, causes, creations, history and a real simple hope for the future.

To my new readers please take understanding to me here. I am not a science mind, guru or savant. I am not putting myself forward as an expert or an EGO presentation of *"No I'm right, No I am, No I said I am" Just* my interpretation of how the grown-ups ACTUALLY discuss things.

My words and thoughts put forward here are for whoever - Scientists, researchers or anyone who may be able to take this baton and run with it - PLEASE DO SO. This is an awakening into my minds visions which may help the world look forward and scale as many hurdles as we can,

ALL IN ONE LEAP.

Chapter One: Just Some Basic Earth Stuff

So here is some simple easy to think about things to help you understand my **Holy Crap Stuff later**. The earth is made up of many different layers. I am not going to bore with them all like Mrs. Jacobs does me, but for easy to understand ideas know this: The earth has layers. The ground which you stand on, the crust, mantle, some sandy layer, a few plates which are like giant flat rocks and all the way in the center we have the core. This is a great big ball of molten lava, gasses, fire and really really hot stuff. Enclosed here I have some simple cut away diagrams (Both Free from the internet as well as an original) of the earth's layers.

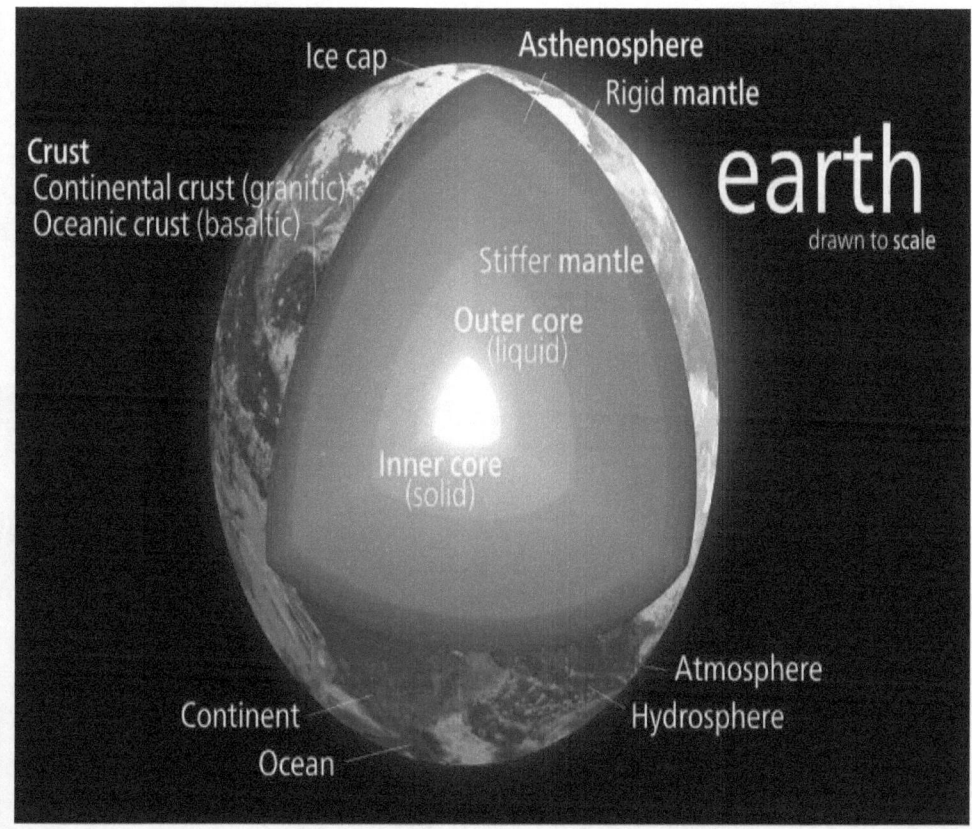

Fig1. Source

https://commons.wikimedia.org/wiki/File:Earth_poster.svg

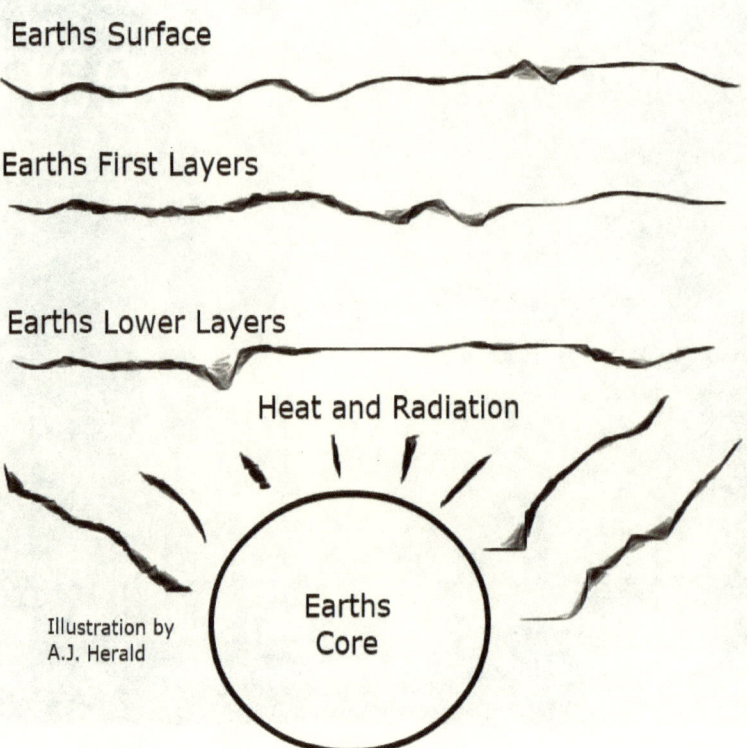

Earths Basics Layers

Earths Surface

Earths First Layers

Earths Lower Layers

Heat and Radiation

Illustration by
A.J. Herald

Earths
Core

My book is written so anyone with basic reading skills

can understand me. I am not being simple because of

thoughts that people are not intelligent or something like that,

but rather I am using what my third grade teacher taught us.
The "KISS method". *"Keep It Simple Sammy"* She use to
tell us that simple is how our minds remember things.
Quotes and sayings, a constant verse in a song - so I am
using this style. Simple.

Why am I talking about the earths layers when writing
a book about Global Warming, *you ask?* Aren't we suppose
to look up at the sky? The sun and ozone layer(s)? A good
question and probably the biggest finding (intuitively)
around. Read each chapter in sync as this tells the story /
beliefs / and my scientific questions in the best possible way.

Simple chapters, simple visions, and hopefully ground
breaking concepts. Scientists, these are for you to have fun
with, ***REALLY!***

Chapter Two: Why The Oil In The Frying Pan?

My mom is the best cook ever. One day I was playing with my dragons at the kitchen table. I had seen my mom at the stove making dinner for that night. She was using the biggest frying pan I have ever seen. It was even bigger than my Brother Joe's head. While using the frying pan, mom had poured in a lot of cooking oil. I saw her doing this and asked, *"Mom why are you doing that?"* *"Doing what A.J.?"* Said mom. *"Pouring in all that oil, like that"*... I said. Mom began to tell me her little secrets of cooking, well until I got older and realized they were everyone's little secrets. Mom began to tell me why she was doing that: *"A.J. when you are cooking on a stove, and you are using a pot or a pan, what happens is the stove's burner does not heat the pan evenly. What tends to happen is the center of the stove's burner gets real hot, while the outer edges of the burner get hot, but not as much. So we use oil to evenly control and dissipate heat over the entire pan. One of the properties of oil is to maintain, control and evenly distribute heat, so there is no one area that gets more heat than the other. If the center of the pan or burner is 400 degrees, then with oil everything is 400 degrees. Nice and even, so the food cooks at the same temperature all together. No food is cooked more than the other."*

Why the heck am I talking about a frying pan and cooking oil in a book about Global Warming? Crazy kid, right? Maybe not, follow me as I easily and simply lead you on.

This way guys to chapter three >>>

Remember Simple

Chapter Three: Dads Wisdom

My dad is a great mechanic. He can fix anything including my sister's broken heart. Daddy really loves what he does even though it doesn't really pay him that much. As he says *"Love what you do and you will never work a day in your life, but you may go hungry from time to time"* Dad teaches me things whenever he can. How to use a wrench and hammer but even more so how to think about things and diagnose problems. He says *"Anyone can swing a hammer, but not everyone can figure out what is wrong with things, even himself from time to time"*.

** This Story Is Important, You Will See **

One day I was helping dad fix the car. While helping him, as usual he gave me some insight and wisdom. He said that *"The art to figuring out problems comes from a person's ability to view and importantly RE-Visualize things from different perceptions. Take a step back, take a step forward or even look at it upside down and somewhere in all of this your mind will see, hear, feel and sense where to go."* He had begun to tell me a story of how he single handedly fixed a car that no one else could, by doing just that. There was a car his shop was working on that had an intermittent problem. Every Thursday, yes Thursday the car owner says while driving to work her car stalls out after making a right hand turn. Ok, so you mean only right hand turns and only on a Thursday? Yes that is what I am telling you. So most mechanics would laugh at this, most did. Some would look at the car to check the basics but no one could find the problem. They all thought the lady was CRAZY, but not my super dad. Dad started by trying to duplicate the problem, yes making right hand turns on a Thursday morning. That was a no go. So then he looked at the basics and everything looked good. After one week dad said to the car owner *"I will come to your house on Thursday morning and you will drive with me in the passenger's seat. I will be in the car with you while the problem happens."* So he did. Guess what, the problem happened, the lady was not crazy. Dad brought the car back to the shop and the guys laughed at him. Dad also laughed, but at himself. Dad says it builds character when a person laughs at themselves. So dad was trying to figure out the problem. He looked at the engine, the gas, the battery and all the stuff that can do this. He found nothing. So while dad was having his lunch he decided to sit down on a little stool he used for certain projects. Usually he eats standing up, but this day his brain was tired so he sat down. He had sat down in front of the ladies car, on the stool and low to the ground. While eating he looked up at the underside of the car's hood. Wallah he saw it, the problem. While sitting down and looking upwards (instead of standing up and looking downwards - LIKE EVERYONE ELSE) he found the problem. He noticed black marks from fire or sparks on the cars under-hood. What had happened was the batteries mounting platform had rotted out on its underside (out of everyone's vision) and the battery was jumping up and arching out on the hood. You cannot see the battery platforms underside as it is hidden from view. The reason why it only happened on Thursdays making right hand turns was that was the day the customer drove to work in a different direction. She would go to the bank on Thursday mornings and cross over a set of railroad tracks enabling the battery to jump up and shut down the car.

The moral, look from all directions. When everyone else is looking down, you look up. What the hell does this have to do with Global Warming? How does this solve the problem or help? Well, while everyone else is looking up, maybe we should be looking down?

Chapter Four: The Big Dig

Ok so getting more into the Global Warming arena. Do we like oil? Do you like oil? Not just cooking oil but oil, oil. You know the stuff the stock market really likes and the same stuff that kills the birds and fishes. That oil stuff. I titled this chapter "The Big Dig "as it has to do with harvesting crude oil. Please don't get me wrong, I like oil or at least what it can do. Drive cars, heat houses, make steel, hey it even helped get man on the moon. What about all the people who love their cell phones but hate oil, hmmm. Do they know that oil put the satellites in space that made those cell phones possible? I know, it's really good stuff just as much as it's really bad. Remember the "KISS" method from my third grade teacher. They would call this one *"A necessary evil"* just to quote Socrates or Confucius or someone very old and smart.

When we dig for oil we drill down into the earth's layers, right? You know first we "Break the Ground" than penetrate the mantle and crust thingy as we go further towards the oil. Hence, The Big Dig (Sorry Boston no puns intended).

Years ago we found oil and what it can do for us. Oh so advanced we have become from the harvesting and refining of this precious resource. No I am not a tree hugger, unless the trees have tits like Mrs. Jacobs, then I am there. I have enclosed a free picture (no not of Mrs. Jacobs) but of an oil drilling rig and cutaway of the earth, just as a visual. I

A.J Herald Registered Copyright
TheGlobalWarmingHerald@gmail.com

would like to take you into my mind where I am going with all of this to see if you can visualize what I am visualizing.

- Remember the story of my dad and looking at things differently?

- Remember the story of my mom and why she used cooking oil?

- Remember the basics of the earth having different layers?

An Original Illustration

Diagram To Follow .. Next Page

A.J Herald Registered Copyright
TheGlobalWarmingHerald@gmail.com

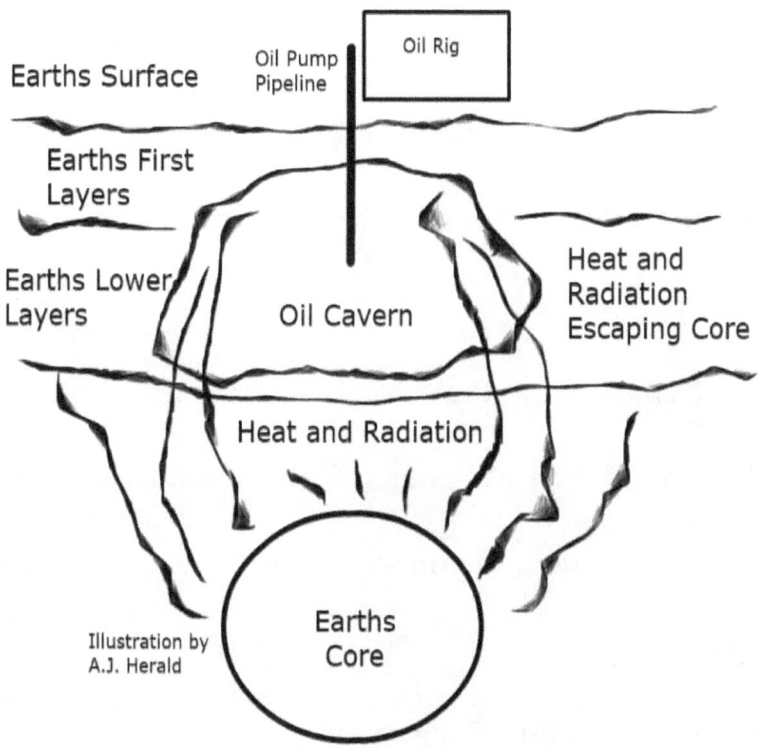

Oil Harvesting Empty Cavern

Illustration by A.J. Herald

Here We Go Into the Meat of It >>>>

Chapter Five: Holy Crap

I mean Holy Crap as a thing mom and dad say sometimes when expressing themselves. I don't mean anything religious.

Remember at the start, **MY DREAM**. Let's take a walk into my dreams vision. A vision which is comprised of the stories I have told you so far, along with a blend of other things.

A Simple Timeline Format:

- It is the year 1950

- Take the layers of the earth and start drilling down through them.

- All the way to the crude oil.

- Now I have struck oil and the economy is going to flourish. Yay!

- Now let's fast forward to the year 2019.

- All that oil I had found in 1950 has been eradicated and harvested for use.

- The world is heating up. We realized that the ice caps are melting, water levels are rising, weather patterns, temperatures and everything else is spazzing out. Is Mother Nature tripping, something dad would say.

- We find that the world's ozone layer in the atmosphere is depleting.

- We form Global Warming conferences, organizations, laws, websites with any and everything we can to help the situation.

- Remember the layers of the earth - especially the earth's core.

- Remember what mom taught me about one of oils properties. To evenly distribute and control heat, **evenly.**

- Remember the story of the crazy lady with the car that shut down. Only On Thursday Mornings. Where everyone was looking in only one direction but dad found the problem by looking another way.

Wait, if oil evenly dissipates and controls heat, and we just syphoned it all from the lower levels of the earth:

- Does that mean that the intense heat from the earth's core is now able to reach the earth's ground layer?

- Doesn't heat exchange and temperature variations create odd and drastic weather patterns? Hurricanes, Tornados, Wild fires etc.

- Isn't it when cold and hot air mix, tornados are formed?

- What about all the oil we found under the sea and ground in Alaska?

- Could the polar ice caps be melting not simply from the sun and ozone, but from the heat dissipating from the earth's core? - Because the oil is not there anymore to protect us from the cores heat.

- Doesn't the earth's core emit radiation as well, not just heat?

- Isn't oil used to cool things from becoming overheated ex. A cooling tower for electricity generators and such?

- What about all the oil from under the ground in Texas and California. Could the California fires be starting because of a dual problem? Lack of water PLUS heat escaping the core? Could the heat escaping the ground be causing the lack of rain, but also increasing the rain elsewhere.

Just as a start into my vision of could be questions, issues and answers. No one could have seen this coming but we can see that it has arrived and now address it, scientifically and rationally.

Chapter Six: My Intuitive Thoughts

As I put together my visions, mom and dad's teachings and my own cognizance, I start to expand on what I am seeing in my mind. If what I have discovered in theory, and my intuitions is true - then can we find simple proof or quasi - proof out there? The following are questions that are open ended and unanswered by design. I wrote these as a triggering mechanism for the minds of all. Like Tony Robbins would say *"The power is in the question"*.

A Simple List of Questions to Ponder:

- Where are all the oil caverns we harvested oil from? If we look at their locations are there changes in weather, specifically in those areas that could help determine this theory?

- Example: Southern California is full of wildfires every year and they get worse. Are there large empty oil caverns underneath that area?

- What about the Gulf area, the area off the coast of Florida etc.? Those are all oil harvesting areas right? Those areas have intensified with weather patterns drastically over the past few years. Is this a coincidence? It just happens to be the oil harvesting areas happen to be in line with bad weather areas? Florida, the Caribbean, Gulf, Southern California, and so on.

- Let's take an area that doesn't have oil harvesting and look at its weather. Let's say Colorado Springs, Montana, Maine, and the Blue Ridge Mountains etc. Or do you think that the elevation of these areas maybe puts more earth between us and the earth's core where the heat cannot escape? Basically those areas are much better insulated. What do you think about this idea? Really, ask the questions to yourself and just contemplate a bit.

- Is the desert in the Middle East becoming hotter because of their oil harvesting?

- Are there ways to see if this theory is true?

- When was the last time we measured the earth's temperatures from below? Every Layer. Has it been recently or are we just utilizing measurements from decades ago?

- Can oil also reduce and protect us from radiation, in this case the earth's radiation?

- If there is a cavern underneath us with oil in it? Does this mean that area is lacking in other layers of the earth that would otherwise be found there? What I mean is if an area without oil has let's say 7 layers of earth, does an area with an oil cavern have only 3 or 4 layers of earth? Meaning the oil makes up the missing few layers of protection? If so what happens when we harvest that oil?

- We have been starring at the Sun and Ozone layer for years. What if we take a look downward? Is there a dual set of circumstances taking place?

- Isn't the Earth's core basically another Sun? If it is just as hot, and made up of the similar elements, then that ball of Sun is a lot closer to us, right?

This book is one of optimism. I am not writing this to bring on dilemmas and stress but help with finding the real problems and, SOLUTIONS. Like dad said *"Anyone can swing a hammer, but you can be the one to diagnose the problems"*

Chapter Seven: My Optimistic Thoughts

If this was the year 2050 and someone said that cellphones have created earth's problems, would we have not had cellphones? Of course we would have them because that is just the way our world works. We invent and produce things like oil and gas all with progress in mind. Yes, of course money is the biggest factor but the reason why things make money is because of us, the consumers and their needs. So with this said I would like to simply state that this is **NOT AN ATTACK ON THE OIL COMPANIES,** as the reason they are here is us. Just like the reason cellphones are here, we the public.

In this chapter I would like to present a whole bunch of optimistic thoughts as to possible answers to help the world eradicate the causes of Global Warming.

A Simple List:

- If harvesting the oil from under the ground aided in the global warming problems, then can putting oil back underground reverse this? What about waste oil, food oil and by products? Can these help with the reversal? If we simply (in theory) replaced the removed oil?

- Here's a grand idea. The oceans are rising in water levels. These levels rising will cause flooding, destruction to our coastlines and possibly making our shores inhabitable. Can we take the rising ocean water and use that, to fill up the oil caverns that have been harvested from the oceans floor? Maybe a containment system that when the water rises to a certain level, it then spills over into the containment apparatus, then piped down into the old oil cavern.

Two problems:

1...How do we fill up an empty oil cavern?

2...How do we lower the rising water levels of the oceans?

Can this 2 for 1 fix-all idea work?

- Now let's talk about the on land oil harvesting. For example in Southern California. If there is an abundance of heat inside of these oil caverns? If we were to pipe in ocean water, would the heat take the water and create steam? If so can that steam be vented through a set of exhaust pipes into the atmosphere? Isn't steam moisture just like fog? Can this steam by any scientific chance turn into humidity and then into rain? Hmmm if this was possible in Southern California, what then?

- Now while on the idea of a 2 for 1 - I would like to talk about tires. Car, truck and rubber tires that is. Do you know how long it takes a tire to decompose? Never, they basically last forever unless you burn or melt them. This is what we do to get rid of tires is burn them, but that in itself now creates burning rubber by-products and fumes. What if we were to take tires and grind them up? Sawdust style. After grinding up the tires we could pipe the small pieces down into the oil caverns where naturally they can melt from the cores heat. When tires melt they will build up a rubber barrier over time creating insulation from the cores heat. Hmmm is this possible?

Conclusion:

All these ideas of mine are as they say "In Theory", but they are possible. My goal from this set of questions and idea triggering is to get the minds of the world, whoever and wherever thinking with a different set of paradigms. Ohh another big word. *"Big words A.J., Big words"*.

"No problem can be solved from the same level of consciousness that created it." - *Albert Einstein*. When Albert said this quote, did he mean that if you have a problem, you need someone smarter to solve it? **NO**, he simply said a different level of consciousness. Operative word **"different"** which may be someone with a much more

simplistic way of thinking. A real life example of this will be this very popular story:

A father and mother are trying to set up their new T.V.'s remote control. With hours of reading the step by step instructions, written in a High School level of English, they still cannot get it to work. Then here comes the 9 year old son and he says *"Daddy let me see that"* and he sets it up within seconds, *"there daddy I got this for you"*. A different but not necessarily smarter level of thinking, or is it?

With love from A.J. Goodnight everybody it's my bedtime now.